Earth

LET'S EXPLORE THE SKY!

By Nicole Horning

Cavendish Square
New York

Published in 2021 by Cavendish Square Publishing, LLC
243 5th Avenue, Suite 136, New York, NY 10016

Copyright © 2021 by Cavendish Square Publishing, LLC

First Edition

No part of this publication may be reproduced, stored in a retrieval system, or transmitted in any form or by any means—electronic, mechanical, photocopying, recording, or otherwise—without the prior permission of the copyright owner. Request for permission should be addressed to Permissions, Cavendish Square Publishing, 243 5th Avenue, Suite 136, New York, NY 10016. Tel (877) 980-4450; fax (877) 980-4454.

Website: cavendishsq.com

This publication represents the opinions and views of the author based on his or her personal experience, knowledge, and research. The information in this book serves as a general guide only. The author and publisher have used their best efforts in preparing this book and disclaim liability rising directly or indirectly from the use and application of this book.

All websites were available and accurate when this book was sent to press.

Cataloging-in-Publication Data

Names: Horning, Nicole.
Title: Let's explore the sky! / Nicole Horning.
Description: New York : Cavendish Square, 2021. | Series: Earth science explorers | Includes index.
Identifiers: ISBN 9781502656339 (pbk.) | ISBN 9781502656353 (library bound) | ISBN 9781502656346 (6 pack) | ISBN 9781502656360 (ebook)
Subjects: LCSH: Astronomy–Juvenile literature.
Classification: LCC QB46.H687 2021 | DDC 520—dc23

Editor: Nicole Horning
Copy Editor: Nathan Heidelberger
Designer: Rachel Rising

The photographs in this book are used by permission and through the courtesy of: Cover ixpert/Shutterstock.com; p. 5 SOPA Images/LightRocket/Getty Images; p. 7 VectorMine/Shutterstock.com; p. 9 Vadim Sadovski/Shutterstock.com; p. 11 Elenamiv/Shutterstock.com; p. 13 Natee Jitthammachai/Shutterstock.com; p. 15 Juergen Faelchle/Shutterstock.com; p. 17 Triff/Shutterstock.com; p. 19 VICTOR TORRES/Shutterstock.com; p. 21 Anton Jankovoy/Shutterstock.com; p. 23 Roman Voloshyn/Shutterstock.com.

Some of the images in this book illustrate individuals who are models. The depictions do not imply actual situations or events.

CPSIA compliance information: Batch #CS20CSQ: For further information contact Cavendish Square Publishing LLC, New York, New York, at 1-877-980-4450.

Printed in the United States of America

CONTENTS

What's in the Sky?	4
Phases of the Moon	12
All About Stars	16
Words to Know	24
Index	24

What's in the Sky?

The sky is above Earth's surface, or ground. The sun, moon, clouds, stars, and **planets** can be seen in the sky. Let's learn about some of the cool things you can see when you look up at the sky!

The sun is seen in the sky during the day. It's very hot! It's made of gases. The sun is at the center of our **solar system**. Earth and the other planets orbit, or move around, the sun.

The sun is much larger than Earth. Large objects in space pull smaller objects toward them. This is how gravity works. The sun's gravity holds our solar system together. It keeps planets moving around the sun.

The sun heats Earth and helps plants grow. Sunlight is made of many colors. When sunlight reaches Earth, gases in the air scatter, or move, the blue part of sunlight most. This makes the sky look blue.

Phases of the Moon

The moon orbits Earth. It looks different at different times of the month. Sometimes it's full. Sometimes it can't be seen at all! This is called a new moon. The different shapes of the moon are called phases.

There are eight phases of the moon. They're caused by the sun. The sun lights up different parts of the moon as it moves around Earth. This makes the eight different shapes.

All About Stars

Many of the little dots of light in the night sky are stars. Stars are often very old. Small stars live longer than large stars. This is because their **mass** is smaller and they're not as hot. They don't burn out as fast.

The sun is a star that's seen during the day. It's called a yellow dwarf star. This means its size is in between very small and very big stars. It's the closest star to Earth. Other stars are very far away.

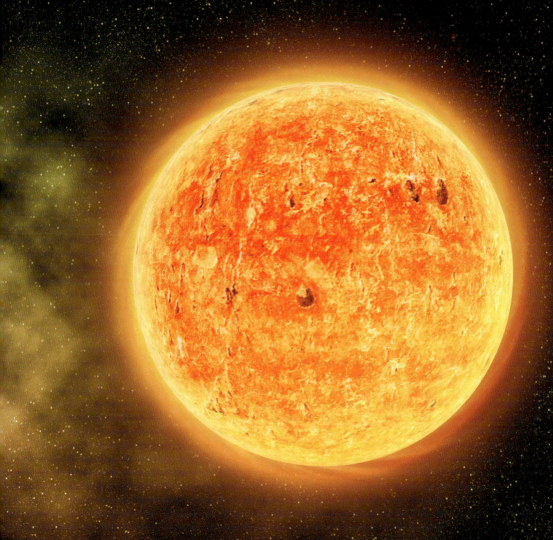

Some stars look brighter than other stars. This can be because of how far away they are from Earth. Stars can also look brighter because of how much **energy** they give off.

Some groups of stars make larger shapes. These shapes are called constellations. People have named constellations after the things they look like, such as fish or bears. It can be fun to look up at the stars and see what shapes you can find!

23

WORDS TO KNOW

energy: The ability to do work.

mass: How much matter is in an object.

planets: Large natural objects, such as Earth, that move around a star.

solar system: The sun and everything that travels around it.

INDEX

C
clouds, 4
color of the sky, 10
constellations, 22

G
gravity, 8

M
moon, 4, 12, 14

P
phases, 12, 14

planets, 4, 6, 8
plants, 10

S
stars, 4, 16, 18, 20, 22
sun, 4, 6, 8, 10, 14, 18